An Illustrated Dictionary of Navajo Landscape Terms

This book is part of the Peter Lang Humanities list.
Every volume is peer reviewed and meets
the highest quality standards for content and production.

PETER LANG
New York • Bern • Berlin
Brussels • Vienna • Oxford • Warsaw

An Illustrated Dictionary of Navajo Landscape Terms

David M. Mark • David Stea • Carmelita Topaha
COMPILERS AND EDITORS

PETER LANG
New York • Bern • Berlin
Brussels • Vienna • Oxford • Warsaw

Library of Congress Cataloging-in-Publication Data

Names: Mark, David M., editor. | Stea, David, editor.
Topaha, Carmelita, editor.
Title: An illustrated dictionary of Navajo landscape terms / David M. Mark,
David Stea, Carmelita Topaha, compilers and editors.
Description: New York: Peter Lang, 2019.
Includes bibliographical references and index.
Identifiers: LCCN 2018044981 | ISBN 978-1-4331-6057-8 (hardback: alk. paper)
ISBN 978-1-4331-6058-5 (paperback: alk. paper) | ISBN 978-1-4331-6059-2 (epdf)
ISBN 978-1-4331-6060-8 (epub) | ISBN 978-1-4331-6061-5 (mobi)
Subjects: LCSH: Geography—Dictionaries.
Geography—Dictionaries—Navajo.
Navajo language—Dictionaries—English.
English language—Dictionaries—Navajo.
Classification: LCC G63 .M375 2019 | DDC 551.403—dc23
LC record available at https://lccn.loc.gov/2018044981
DOI 10.3726/b14563

Bibliographic information published by **Die Deutsche Nationalbibliothek**.
Die Deutsche Nationalbibliothek lists this publication in the "Deutsche
Nationalbibliografie"; detailed bibliographic data are available
on the Internet at http://dnb.d-nb.de/.

The paper in this book meets the guidelines for permanence and durability
of the Committee on Production Guidelines for Book Longevity
of the Council of Library Resources.

© 2019 Peter Lang Publishing, Inc., New York
29 Broadway, 18th floor, New York, NY 10006
www.peterlang.com

All rights reserved.
Reprint or reproduction, even partially, in all forms such as microfilm,
xerography, microfiche, microcard, and offset strictly prohibited.

Printed in the United States of America

Contents

List of Figures.. vii
Saad Ałtsé Si'ánígíí (Preface)..ix
Project History xi
A Note on the Methodology xiii
Acknowledgments... xv
The Organization of This Book xvii

Section 1: Water-related Features..................................... 1
Section 2: Elongated Depressions................................... 19
Section 3: Open Spaces, Gaps, and Holes 31
Section 4: Elevations and Rock Formations.......................... 47
Section 5: World, Land, Place...................................... 59
Section 6: Vegetation 69
Section 7: Earth Materials... 75

Index to Navajo-language Terms 85
Index to English-language Terms 91

Figures

Note: All photographs taken by David M. Mark or are from Wikimedia or the U.S. Government; Photographs © David M. Mark 2018

Tooh, the San Juan River near Mexican Hat, Utah 3
Tó nílį́ in Canyon de Chelly 5
The open ocean, viewed from shore 7
'Adahiilį́ or **'Adahiilíní (Grand Falls** in English) on the Little
    Colorado River, an example of **tó hadah dadeezlį́**.
    Source of photo: Wikipedia 9
Be'ek'id in the Chuska Mountains 11
An example of a **ch'ínílį́** near Fruitland, New Mexico 13
Tééh, a natural pool on the flat rocks above Canyon de Chelly 15
An example of **nahodits'ǫ'** near Sanostee 17
Example of **dah nahwii 'eeł** near The Hogback, San Juan Chapter 17
Chaco Wash, New Mexico, near where it joins the San Juan River 21
A smaller **bikooh**, Tóhdildoní Wash, just northeast of the town of
    Navajo, New Mexico 21
Little Colorado River Canyon, Arizona 23
Crow Canyon, a side canyon off Largo Canyon in Dinétah, the
    Navajo homeland 23
Canyon de Chelly, Arizona 25
A small "dry wash" in Dinétah, off Largo Canyon 27
Nástł'ah at the head of a side canyon in Upper Fruitland 29

Biníí' dah hastł'ah refers to more than one cove. The formation shown here is just south of Window Rock, Arizona 29
halgai, just south of Shiprock pinnacle, looking south from the road to Cove 33
Open country along Highway 264, east of Tuba City, Arizona 35
The horizon over flat open land 37
A gap in a ridge just west of Lybrook, New Mexico, is an example of **bigiizh** 39
The gap behind this rock at Chaco Canyon would be referred to as **tsék'iiz** 41
'Adah Hosh Łání (Meteor Crater, Arizona). Source: Wikipedia 43
'a'áán, a cave in lava rock at El Malpais, New Mexico 45
Tsoodził, also known as "Mount Taylor" in English, is the Navajo Sacred Mountain of the South 49
Dah násk'id 51
Dayílk'id 51
Séí yáalk'id 52
Bis dah 'azką́ (Table Mesa), just south of Shiprock, New Mexico 53
Tsé łichíí' dah 'azkání (Red Mesa) in Red Valley, Arizona 53
Tsézhin 'íí'áhí, known as "The Thumb" in English (Red Valley, New Mexico) 55
Tséghahoodzání: The rock arch at Window Rock 57
Near Shiprock Pinnacle, New Mexico 58
Earth. Source: NASA 61
Diné Bikéyah, "Navajo Country," the Land of the People. Source: Wikipedia 63
Una Vida 'Ruins' at Chaco Culture Park, New Mexico 65
Tsé náázhoozh, "a rockslide happened," at the north end of Table Mesa, New Mexico 67
Hootso near Red Valley, Arizona 71
Piñon-Juniper woodland 73
A boulder, a lump of rock (stone) 77
séí dah daask'id (sand dune) 79
bis 81
hashtł'ish 81
Picture taken near an old cabin in Red Valley, Arizona 83

Saad Ałtsé Si'ánígíí (Preface)

Díí ashdladiin kéyah dah si'ánígíí biyi' bitsi' yishtłízhii kéédahat'ínígíí Diné bi'di'nínígíí éí ts'ídá aláahdi t'áá bí bizaad yee yádaałti'. Ákonidi hastóí dóó sáanii t'áá bí bizaad yee yádaałti'ígíí hadaastih dóó nihits'ą́ą' bihidínídééh daazlį́į́' biniinaa Diné bizaad t'áá íiyisí yaa kódzaa háálá ánii háánoot'áanii doo t'áá bí bizaad yee yádaałti' da dóó doo yídahooł'aah da. Díí t'áá aanií ákót'éego biniinaa kéyah ał'ąą át'éego nidahaz'ánígíí dabízhi' doo bił béédahózin da. Díí kéyah ał'ąą ádaat'éego bił nidahaz'ánígíí bízhi' dahólǫ́ǫgo chodaa'ínígíí t'áá íiyisí k'ad ánii háánoot'áanii doo bił béédahózin da dóó doo chodayooł'įį da háálá Diné danilínígíí lą́'í kin hadaas'áágóó dóó kintahgóó adahaaznáago k'ad áadi kéédahat'į́. Diné bikéyah yits'ą́ąjį' adahaaznánígíí Diné bikéyah bikáa'gi dahózhónígo nidahaz'ánígíí kéyah ádaat'éegi yaa yádadoołtih yę́ę doo bił béédahózin da daazlį́į́' dóó dayózhíigo ádaalyéhígíí dóó ha'át'íí biniiyé ákót'éego ádaasye'ígíí doo bił béédahózin da dóó doo yaa yádaałti' da áko éí ał'ąą át'éego kéyah dabízhi'ígíí ałtso baa hoyoo'nééh dóó baa hoyoosnah.

Díí kót'éego kéyah ał'ąą ádaat'éego nidahaz'ánígíí dóó dabízhi'go ádaolyéhígíí éí t'áá íiyisí aláahgo baa nitsáhákees dóó ha'át'íí biniiyé ákódaosye'ígíí Diné bizaad dóó bee é'ool'įįł dóó binahjį' nát'ą́ą' náháne' bii' naaznil ako íl'įįgo baa nitsáhákees. Díí naaltsoos hadilyaaígíí éí kót'éego kéyah ał'ąą ádaat'éego nidahaz'ánígíí aheeskid dóó hane' yázhí bídaoltą'go ályaa ako náás hodeeshzhiizhgóó doo baa dahodiyoo'nah da biniiyé dóó díige'át'éego naaltsoos ályaago éí doo nanitł'agóó bídahoo'aah dooleeł ílį́įgo dóó da'ólta'góó da chodayooł'įį dooleeł biniiyé dóó t'áá hooghan góne' da díí bínídahoo'aah dooleeł biniiyé naaltsoos hadilyaa.

Project History

The research project that produced this *Illustrated Dictionary of Navajo Landscape Terms* began more than 10 years ago and some 10,000 miles away from Navajo country. The dictionary project, while in process for nearly a decade, is not complete, since the Navajo language is a living language that changes with time, history, and the lived experience of the various members of the Navajo Nation. We anticipate that the current version of the Dictionary will be both interesting and useful to the Navajo people in general, and especially to students in schools and language classes. We would greatly appreciate receiving any comments or corrections that the reader may choose to offer, and our contact address is below.

The history of the project spans the globe. In October 2002, Professor David Mark went to Western Australia where Professor Andrew Turk lives, to begin a project whose objective is the documentation of landscape terms and concepts of the Yindjibarndi language, currently spoken by a few hundred people living in the Pilbara region of Western Australia. In May 2003, Professor Mark presented some results from the Yindjibarndi project to the Geography Department at Texas State University in San Marcos. After the lecture, Professor David Stea and a few others in the audience pointed out that the landscape of Yindjibarndi country resembles some parts of the Navajo Nation. Professors Mark and Stea visited Navajo country in July 2003, and decided to attempt a comparative study between the landscape categories and concepts present in the Navajo and Yindjibarndi languages. Mark and Stea applied for a grant from the US National Science Foundation to support the project, and the grant was awarded starting in September 2004. We received a research permit from the Cultural Resource Compliance Section of the Navajo

Nation Historic Preservation Department, permitting us to conduct landscape and language research on the Navajo Reservation. Over the following six years, Mark and Stea made more than a dozen trips to Navajo country to interview monolingual and bilingual speakers of the Navajo language and to obtain advice on the project. One of the main results of the project is this Dictionary.

David M. Mark
Department of Geography
University at Buffalo
Buffalo, New York USA 14261

or send email to:

dmark@buffalo.edu

A Note on the Methodology

We conducted many "field interviews" in which we drove around the Chapter (local government) areas of Navajo speakers and asked them to tell us what Navajo terms they would use for what we were seeing. We also took many photographs during those driving sessions and indexed the locations of the photos with the aid of a GPS (Global Positioning System) device. We also arranged ten sessions in which small groups of Navajo speakers were assembled in a room and while looking at landscape photographs taken on or near the Navajo Reservation, told us what they saw in the photographs. The drive-around and photo-response sessions were conducted over much of the reservation, from Ojo Encino to Shonto and many places in between. In several other sessions, held in Farmington, Navajo-speakers were asked to divide some landscape photographs into groups based on similarity; they were then asked to say what it was that the photos within a group had in common. Finally, we looked up English-language landscape terms in Young and Morgan's English-Navajo dictionary to find other Navajo landscape terms. "A Navajo-English Thesaurus of Geological terms" by Blackhorse, Semken, and Charley (2003) was also found to be valuable.

From all of these sources of information, we compiled a draft version of the Dictionary, and selected photographs to illustrate the meanings of as many of the Navajo terms as possible. Then we reviewed the draft with expert speakers of Navajo, who suggested additions, deletions, and corrections, and gave advice on which photographs best illustrated each term; at the request of some of our consultants, we have omitted from this dictionary some old words for landscape features that now are only used in traditional origin stories and prayers. Toward the end of the project, we asked language experts to take us to locations where we could photograph additional landscape features to illustrate the remaining terms. Then, the complete draft was edited by Navajo language expert Irene Silentman, who made final corrections, mostly in spelling and grammar but also regarding the meanings of some of the terms. Any errors that remain in this document are the responsibility of Mark and Stea, and we welcome comments and corrections.

Acknowledgments

First and foremost, we wish to thank all the speakers of the Navajo language that we have met, too numerous to mention here, who shared their knowledge of the Navajo language, culture, and landscape with us. The Cultural Resource Compliance Section of the Navajo Nation Historic Preservation Department issued Permit C0513-E, permitting us to conduct landscape research on the Navajo Reservation. Completion of the project would not have been possible without the tireless assistance of Carmelita Topaha, who is a co-author of this dictionary. We also wish to thank the U.S. National Science Foundation for providing most of the funding required to support this project, in the form of grants BCS-0423075 and BCS-0423023. Irene Silentman provided valuable and detailed editing of a complete draft of this dictionary, correcting the spelling, translating the preface, and providing other valuable advice. Larry King, Harding Yazzie, Jr., and Leonard Dan also provided important input. We also received valuable advice and input from Professor Andrew Turk and from Dr. Jay Williams, University of New Mexico.

About 50 people, mostly Navajo Elders, participated in our research. In alphabetical order, these included: Betty Anderson, Irene Arviso, Harry Barlow (Shonto), Erma Batteso (Farmingon), Lucy Bee, Markie Bee, Charlie Begay (Piñon), Daisy Begay (Sanostee), Helen Begay (Piñon), James M. Begay (Round Rock), Lula Begay (Sanostee), Mae Ann Begay (Piñon), Rosemary Begay (Piñon), Sadie Benally (Piñon), Sarah Benally (Kirtland), Elaine Benaly (Farmingon), Irene Bennalley, Mabel Yazzie Bennalley, Deborah J. Bennett, Irene Bennett, Barbara Billy (Burnham), Robert Black, Jr. (Shonto), Stephanie Bowman (Newcomb), Claudia Clitso (Shonto), Robert Clitso (Shonto), Amelie Collins (Farmingon), Charlotte

Collins (Farmingon), Art Enoah (Newcomb), Julia Enoah (Newcomb), Nia Francisco (Navajo), Freda Garnañez (Shiprock), Raechel Halwood, Jennie Harvey (Round Rock), Benjamin Hogue, Lolita Hogue, Loretta Holyan (Fort Defiance), Louise Hubbell (Farmington), Angelita Joe (Lukachukai), Lorraine John (Newcomb), Serena Jumbo (Sanostee), Dolly Kaye (Piñon), Larry King (Red Valley), Bessie Lansing (Sanostee), Mae Lewis (Sanostee), Ted Mace (Ojo Encino), Lorraine Manavi (Farmington), Bessie Nez (Shonto), Keo Scott (Piñon), Peter Topahe (Red Mesa), Nancy Tso, Henry Tsosie, Rena Tsosie (Piñon), Daisy Yazzie, Grace Yazzie (Sanostee), Harding Yazzie, Jr. (Farmingon), Irene Yazzie (Piñon), and Rosie Yazzie (Shonto). A few other participants wished to remain anonymous. We apologize to any who participated and who, through our own oversight have been omitted from the list above.

The Organization of This Book

Most dictionaries and encyclopedias are arranged alphabetically by terms. However, we have decided to organize this dictionary by the similarity of the topics (subject matter). We hope that readers will enjoy browsing through the sections that contain terms for similar features.

In Section 1, we present Navajo terms for water-related landscape features. Section 2 includes Navajo terms for elongated depressions in the landscape that might be referred to in English by terms like canyon, gully, or dry wash. Section 3 includes terms for broad level area and for other concave forms such as gaps and holes. Section 4 presents Navajo terms for mountains and hills, buttes, mesas, and other rock formations. Sections 5, 6, and 7 cover terms for earth and land, for vegetation and for earth materials, respectively.

For those readers who want to look up particular terms, we have added two indexes at the end. The first one lists Navajo-language terms in alphabetical order, with page references for each. The second index gives English-language terms and their page numbers.

Section 1
Water-related Features

tooh or tó

English words with similar meaning:

body of water

Tooh, the San Juan River near Mexican Hat, Utah

Tó and **tooh** are both words that mean the same as "water" in English. **Tooh** is used in terms for relatively larger bodies of flowing water, and **tó** for relatively smaller water features, whether the water is flowing or still.

Some related terms for water in the environment that is not flowing:

tó dah siyí
　　a lake or pond.

tó siyínígíí
　　a lake, a body of water sitting. **Tooh siyínígíí** refers to a large body of water, such as Wheatfields Lake [near Wheatfields, Arizona]. Refers to a specific place where water is sitting.

tó biníhííyí
　　an inlet, bay, or harbor in the shore of a lake, river, or stream.

tábąąh
　　the shore, beach, water's edge, bank (of a lake, river, or stream). This term is the source of a Navajo clan name.

tó dah néigeeh or **tó dah néigeehígíí**
　　periodic, sometimes dry, water body used for stock watering. A place where water gathers.

tó nílį́

English words with similar meaning:

stream

Tó nílį́ in Canyon de Chelly

Tó nílį́ refers to a stream of running water.

Some related terms for flowing water in the environment:

tooh
    a large body of water, flowing, a big river. **Tó tsoh** is another way to refer to a large body of water.

tó nílínígíí
    a stream, running water. Larger ones (rivers) might be referred to as **tooh nílínígíí**.

tó ahidahdiilį́ or **tó 'ahidahdiilį́**
    a tributary or side stream.

tó biyáázh
    a small spring (of the type generated by rain or melting snow); spring water, literally, **biyázhí** refers to offspring or child.

tó yilą́ąh
    "flood comes"; literally, "water becoming more". Increasing water.

The open ocean, viewed from shore

tónteel

English words with similar meaning:

ocean, sea, very large lake

Tónteel refers to the ocean such as the Pacific. Young and Morgan's dictionary states that this term also would be used for a very large lake, such as Great Salt Lake, but some current speakers say it only can refer to the ocean.

Related term:

tónteel bibąąh
    seashore, beach

'Adahiilį́ or 'Adahiilíní (Grand Falls in English) on the Little Colorado River, an example of **tó hadah dadeezlį́**.
Source of photo: Wikipedia[1]

tó dah 'iilį́
or
tó adah 'iilį́

English words with similar meaning:

waterfall

tó adah 'iilį́ is the general term for a waterfall.

Dah 'iilį́ on its own refers to any fluid material flowing or falling down from an elevation. It is usually combined with **tó** for terms meaning "waterfall".

---

1 http://en.wikipedia.org/wiki/File:Grand_Falls_of_the_Little_Colorado_River_near_Flagstaff,_Arizona.jpg

Related terms:

tó hadah 'iilį́
  term for a waterfall when you are standing at the bottom looking up

tó bidah 'iilį́
  waterfall

Some other terms for water features that would be called "waterfall" in English:

- **tó hideezlį́** not a fall, but a trickle down the face of the rock
- **tó ndeezbaal** a near-vertical cascade over rocks, a curtain of water, you could be behind it; **baal** means a hanging curtain

Be'ek'id in the Chuska Mountains

be'ek'id

English words with similar meaning:

small lake, natural pond

be'ek'id
Water sitting still. The term usually refers to a small lake or natural pond. A **be'ek'id** can be in the lowlands or up in the mountains. It is still **be'ek'id** even if it has no water in it.

An example of a **ch'ínílį́** near Fruitland, New Mexico

ch'ínílį́

English words with similar meaning:

water outlet

Ch'ínílį́ refers to a location where water runs out of some place, such as a canyon. This term applies to the location, even when the water is not flowing.

Chinle, Arizona, the town at the outlet of Canyon de Chelly, is an English-language spelling of **Ch'ínílį́**, its Navajo name.

tééh

English words with similar meaning:

natural water basin

Tééh, a natural pool on the flat rocks above Canyon de Chelly

Tééh usually refers to a natural basin that can hold water, a natural cistern.

Young and Morgan[2] list two additional meanings for tééh: "valley, deep water"

Tééh can refer to a deep pit with water. In the Newcomb area at least, people use the term to refer to Chaco Wash. In this case **tééh** is the wash itself; with the water running they say "**tééh nílį́**." Chaco Wash is a deep valley that runs along from south to north into the San Juan River. People around Newcomb may say that their cows are at **tééh**.

---

2   Young and Morgan, p. 994.

nahodits'ǫ'

An example of **nahodits'ǫ'** near Sanostee

English words with similar meaning:

mudhole, bog

Nahodits'ǫ' refers to a mudhole or bog; it is a sticky place where a person attempting to cross could get stuck in mud or quicksand.[3]

Example of **dah nahwii 'eeł** near The Hogback, San Juan Chapter

---

[3] Our Navajo consultants used the English word "quicksand", even though the material did not seem to meet the usual English-language definition of "quicksand".

Related terms:

dah nahwii 'eeł is a term for a bog or marsh, covered with water, with a muddy bottom into which one's feet might sink a short distance.

hashtł'ish dits'idí
a mudhole where more force would be needed to get out of it than from a more usual mudhole of this type. Sticky clay. When it gets deep, it would be called **nahodits'ǫ'**.

Section 2
Elongated Depressions

bikooh

English words with similar meaning:

canyon, arroyo, gully, ravine

Chaco Wash, New Mexico, near where it joins the San Juan River

A smaller **bikooh**, Tóhdildoní Wash, just northeast of the town of Navajo, New Mexico

The Navajo word "**bikooh**" is used to talk about a variety of landforms that would require several different words in English. A **bikooh** is a long, sloping depression in the ground, meaning that water would flow along this depression or natural trench after heavy rains or snow melt. Sometimes, **bikooh** refers to large valleys with steep sides, that would be called canyons in English. But some **bikooh** are much smaller and would be called washes or arroyos in English. If the sides of the feature are mostly exposed bedrock,[4] then **tsékooh** is used instead (see "**tsékooh**"). Smaller steep valleys, that might be called gullies or gulches in English, also could be called **bikooh** (or **tsékooh**) in Navajo—**bikooh** can refer to long, concave features of almost any size, if water could flow along them after a rain storm.

---

4   We are using this term "bedrock" to refer to solid rock that is part of the Earth's crust, whether it is exposed horizontally or vertically. The term "bedrock" is to distinguish this from detached rocks such as boulders or pebbles.

In New Mexico, the word **bikooh** also refers to a dry wash or arroyo, a stream bed that is usually dry but where water flows after rains. A **bikooh** of this type can run along the floor of a **bikooh** that is a valley.

In Arizona, the term **bikooh** is used for large longitudinal depressions (that would be called canyons or gullies in English), but washes and arroyos (dry stream beds) are usually called **cháshk'eh** in Western Navajo. (see "cháshk'eh")

Related terms:

Modifiers can be added to **bikooh** to refer to a big canyon or wash (**hatsoh**), or a narrow one (**hats'ózí**), etc. These are old terms that are seldom used now.

Other old and seldom used terms include: **'abikooh,** a canyon that extends away out of sight, perhaps into a dense forest; **'ahéébíkooh,** a curved canyon that forms a partial circle (from a perspective down in the canyon); **naanázkooh** is a term for a curved canyon as seen from the rim of the canyon; **'ahibidiikooh,** a canyon that converges or joins another canyon downstream, a tributary or side canyon; **'ałts'ábíkooh** divergent or forked canyons, looking upstream; **ch'íbíkooh** an emerging canyon; and **bikooh bidáá'** an edge of the rim of a canyon or arroyo.

Little Colorado River Canyon, Arizona

tsékooh

English words with similar meaning:

rock canyon

Tsékooh refers to a longitudinal depression in the Earth, that a river or creek sometimes or always runs through, and that has sides made primarily of exposed bedrock.

Tsékooh bidáá' refers to the edge, brink or rim of a rock canyon, as shown in the foreground of the picture above.

Crow Canyon, a side canyon off Largo Canyon in Dinétah, the Navajo homeland

Crow Canyon (above) seems to be a borderline case between **tsékooh** and **bikooh**. Some Navajo language experts pointed out that the sides of the canyon are partially covered with loose rock, soil, and vegetation, and thus the term bikooh should be used. Others said that there is enough exposed bedrock in the walls for the word **tsékooh** to be used.

Canyon de Chelly, Arizona

tséyi'

English words with similar meaning:

rock canyon

Large elongated depressions in the earth, with sides composed of rock, are sometimes called **tséyi'** in Navajo. **Tséyi'** literally means "within the rock": tsé = rock, -yi' = within or between, thus "rock canyon". This term means about the same as **tsékooh**.

The word **tséyi'** is the source of the English-language names of some canyons in Arizona, including Canyon de Chelly and Tsegi Canyon, near Kayenta, Arizona; "Chelly" is just **tséyi'** after being written down, first in Spanish and later in English.

cháshk'eh

English words with similar meaning:

dry wash, gully, arroyo, stream bed

A small "dry wash" in Dinétah, off Largo Canyon

Cháshk'eh is a Navajo word for a dry wash or arroyo, a sandy stream bed that is usually dry. The term is used mainly in Arizona (Western Navajo). People in New Mexico know this word, but tell us that it is used in Arizona but not in New Mexico, since in New Mexico those features would be called **bikooh**. However, **cháshk'eh** does not seem to be used anywhere for the "canyon" sense of **bikooh**; in Arizona, large features that would be called "canyons" in English are called **bikooh, tsékooh,** or **tséyi'**, just like they are in New Mexico; only washes or arroyos are **cháshk'eh**.

As in the case of **bikooh, cháshk'eh** can refer to a large wash (**hatsoh**) or a narrow one (**hats'ózí**). **Cháshk'eh bidáá'** is the term for the edge of a dry wash.

(See also -k'eh, 'place' page 65)

nástł'ah

English words with similar meaning:

cove[5]

Nástł'ah at the head of a side canyon in Upper Fruitland

The term **nástł'ah** often refers to an indentation into a cliff. It also can refer to an inside corner or curved recess inside a building. Seen from above, a typical **nástł'ah** is not as long as a canyon (**bikooh, tsékooh**). Most **nástł'ah** have a vertical or near vertical cliff at their upstream ends. Some of them are "dead-end" canyons that might be called "box canyons" or "blind canyons" in English.

Biníí' dah hastł'ah refers to more than one cove. The formation shown here is just south of Window Rock, Arizona

---

5 In standard English, the word "cove" normally refers to a small bay in the shoreline of a water body. But in some versions of English "cove" refers to an indentation into the side of a hill.

Biníí dastł'ah or **biníí dah hastł'ah** refers to more than one indentation.

Tséstł'ah haazt'i' refers to features that are similar to **nástł'ah**, but must be deep enough to form a possible trap for an animal or person, while a **nástł'ah** could be only slightly indented and does not need to be narrow.

Tsístł'ah haazt'i' could refer to an indentation in the landscape that was made by people for use as a trap.

Section 3
Open Spaces, Gaps, and Holes

halgai, just south of Shiprock pinnacle, looking south from the road to Cove

halgai

English words with similar meaning:

plain, flatland, valley

halgai
    a vast plain, flat land, a valley with little or no vegetation.

dadzígai
    valley or vast plain; transition from a hill to a valley (eg. Largo Canyon); plain on the bottom of a canyon.

'adzoogai
    valley or plain; this term can refer to a long, open path cleared by people, such as for a power line.

dzigai
    with a non-high "i" (**dzigai**). this term refers to the land at the horizon; with the high-tone í (**dzígai**), this term refers to the sky just above the horizon.

ndzisgai
    valley or plain; rolling country; flat in different directions, with little valleys, such as in Bisti badlands.

Open country along Highway 264, east of Tuba City, Arizona

-teel[6]

English words with similar meaning:

open, flat area

hazteel
    a wide flat area

hóteel
    a large flat area that is wide. This term implies wide open area.

honíteel/hońteel
    a large flat area that is long. This term implies long distances.

náhodékaad
    flat open area, especially one with flattened vegetation, or vegetation in the process of restoration or re-growth, such as after a wildfire.

---

[6] See also **tónteel**, ocean or large lake, page 7.

ni'nineel'ą́

English words with similar meaning:

horizon

The horizon over flat open land

ni' nineel'ą́
    the earth's end, the far horizon

yá nineel'ą́
    horizon, skyline

Ni' refers to the ground and **yá** to the sky. The horizon is, in a sense, where the ground meets the sky, and it can be expressed as the end of the ground or the end of the sky.

The horizon behind mountains or on water or behind the land would be denoted by **dził nineel'ą́**, **tó nineel'ą́**, and **kéyah nineel'ą́**, respectively.

bigiizh

English words with similar meaning:

gap; saddle; pass; cut

A gap in a ridge just west of Lybrook, New Mexico, is an example of **bigiizh**

The stem **-giizh** refers to a slender cut. It can refer to a narrow crevice in the rock. Combined with **dził** ("mountain"), **dziłgiizh** refers to a mountain pass, or to a gap in a mountain range. **Tségiizh** is a cut in the rocks.

-k'iz

English words with similar meaning:

crack, crevice, narrow gap

The gap behind this rock at Chaco Canyon would be referred to as **tsék'iiz**

tsék'iiz is a gap between rocks, refers to a cleft in the rocks.

-niik'iiz refers to a cleft, crack, or gap, such as one that can be used by a person to climb cliffs or rocks.

yisk'iz means a clean crack, a hairline fracture.

Note: **k'iiz** and **k'iz** have about the same meaning, except that **k'iiz** is perfective mode (similar to past tense) and **k'iz** is imperfective (similar to present).

hahoots'aa'

English words with similar meaning:

crater

'Adah Hosh Łání (Meteor Crater, Arizona).
Source: Wikipedia[8]

This term is listed in *A Navajo-English Thesaurus of Geological Terms*[8] as meaning "crater". It can apply to any kind of closed depression in the Earth's surface.

Háádahazts'aa' is the plural form, but more literally means something having a large number of holes, or "pock-marked."

---

7   http://upload.wikimedia.org/wikipedia/commons/c/cf/Meteor.jpg
8   Blackhorse, A., Semken, S., and Charley, P., 2003. *A Navajo-English Thesaurus of Geological Terms*. New Mexico Geological Society and Diné College, Shiprock, Navajo Nation, New Mexico.

[some other words for holes and depressions in the Earth]

'a'áán, a cave in lava rock at El Malpais, New Mexico

Here are some additional Navajo terms for holes or depressions in the Earth's surface.

'a'áán
    hole, burrow, tunnel, etc., in the ground or into a mountain. A natural cave. This term can also refer to small holes in the earth such as Prairie Dog holes.

hoolts'aa'
    a large, concave depression in a surface (such as on the top of a rock or in the ground), an open, flat bowl shape. The term can also refer to the bed of an intermittent lake. **Náhoolts'aa'** is a wide, shallow depression.

náhooldzis
    another word for a concave depression in the ground, a basin (concave depression), deeper than **hoolts'aa'**, not wide

Section 4
Elevations and Rock Formations

Tsoodził, also known as "Mount Taylor" in English, is the Navajo Sacred Mountain of the South

dził

English words with similar meaning:

mountain

The Navajo word **dził** refers to a large feature of the landscape that stands high above its surroundings. The English word "mountain" is usually a pretty good translation. But some of the features in Navajo country that are called **dził** in Navajo are not called "mountain" in English. Examples are Black Mesa (**Dził Yíjiin** in Navajo) and the Defiance Plateau (usually just **Dził** in Navajo).

Related terms:

dził dah 'ats'os: peaks

dził látah: summit (exact top) of a mountain

dziłgháá': mountaintop, summit; anywhere on the mountain

dził ní'á: mountain range

dziłtł'ah: mountain recess, pocket in a mountain, base of a mountain

dziłtát'ah: ledge or shelf on the side of a mountain, mountainside or slope

dziłniit'aa: side of a mountain slope, foot of a mountain

dziłbąąh: mountainside; anywhere on the mountain, with lots of trees and plants.

dził binánii: mountain slope, hillside; also a meadow on the mountain, half way up.

dził diyinígíí: one of the sacred mountains; **dził dadiyinígíí** means several sacred mountains

Dah násk'id

dah násk'id

English words with similar meaning:

hill

Dah násk'id refers to one hill.

Dayílk'id refers to more than one hill.

Dayílk'id

Dah yisk'id means "there is a hill, mound, or promontory," **haalk'id** means "there is a mound or knoll," and **haask'id** means "a hill coming toward me, peaking up."

Séí yáalk'id

Séí yáalk'id refers to a sand hill or sand dune.

dah 'azką

English words with similar meaning:

mesa, plateau

Bis dah 'azką (Table Mesa), just south of Shiprock, New Mexico

dah means "at an elevation". Young and Morgan (1987, p. 300) wrote that **'azką** means "flat-topped".

tsé dah 'azką
   rock mesa

Tsé łichíí' dah 'azkání (Red Mesa) in Red Valley, Arizona

tsé 'íí'áhí

English words with similar meaning:

Rock standing vertically

Tsézhin 'íí'áhí, known as "The Thumb" in English (Red Valley, New Mexico)

tsé 'íí'áhí
    monolith, rock spire

The color and/or material can be described, as in the example above, where **tsézhin** means dark lava rock. The shape of the top also can be described: **bikáá' dah nímaz** would indicate a rounded top, a dome. There are also terms for a jagged top, or a pointed top.

tsé 'íí'á
    pinnacle. This term refers to a single rock spire or butte.

tsézhin 'íí'áhí
    lava spire, volcanic core

Chézhin and **tsézhin** have the same basic meaning. The spelling is different to indicate different pronunciation. **Chézhin** may be more common in Western Navajo. But also, **chézhin** is definitely lava rock, whereas **tsézhin** just means "black rock" or "dark rock".

tsé

English words with similar meaning:

rock

Tségháhoodzání: The rock arch at Window Rock

In this entry, we list some Navajo terms for rock outcrops. The Navajo language does not include a term for a rock outcrop in general, "you have to describe it."

Some common compound terms starting with **tsé** were listed earlier in this dictionary. These include **tsékooh, tséyi', tsé dah 'azką, tsé 'íí'áhí**, and **tsé náázhoozh**

tsé si'ą: rock, sitting. This term could refer to a butte that is neither "standing" nor "lying down".

tsé biih hoodzáán: a general term for a hole in the rock, perforated rock, rock arch.

tsé nané'á is the term for a natural bridge or arch.

tsé dah 'ats'os: a pointed rock or peak; "**dah 'ats'os**" refers to a pointed feature.

tsé yík'áán is the Navajo term for a sharp ridge formed from sloping rocks. In English, these formations are often called "hogbacks". **Tsé yík'áán** is the name of several features in Navajo country.

(rock outcrop terms, continued)

tsé yaa hatsoh: rock (overhanging ledge), a big hollow under a rock overhang; could be occupied by cliff dwellings such as at Mesa Verde.

Near Shiprock Pinnacle, New Mexico

tsé níti'í: vertical rock formation that is like a fence.

tsé k'aal: a rock trap (a gap in a rock that hunters could close to trap animals that they were hunting). This seems to be an old word, now used mainly in ceremonies.

Section 5
World, Land, Place

nahasdzáán

English words with similar meaning:

the world,
the Earth

Earth. Source: NASA

Nahasdzáán: the world, the Earth World, Mother Earth.

Related terms:

ni'
the surface of the earth, the ground; "downward". This is a spiritual term, probably used mainly in Western Navajo.

nihosdzáán
on Mother earth, on the ground, in the world, the Earth World

ni' bikáá'
on the Earth (another way of saying "nihosdzáán")

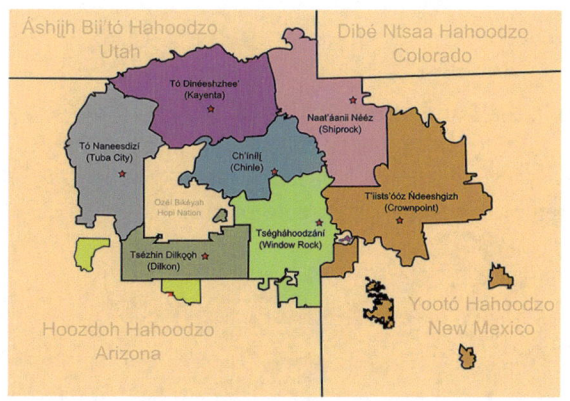

Diné Bikéyah, "Navajo Country," the Land of the People. Source: Wikipedia[9]

kéyah

English words with similar meaning:

land,
real property,
homeland

Kéyah refers to land, real property, including homeland. It also appears in related compound terms and phrases.

---

9 Source of this image: http://nv.wikipedia.org/wiki/Diné_Bikéyah_be'elyaa%C3%ADg%C3%AD%C3%AD/Diné_Bikéyah_bisiláo_dóó_be'atiin_yaa_áhályán%C3%AD_(nahós'a')

Related terms:

Diné Bikéyah, "Navajo Country", the Land of the People

shikéyah
"my land", as in a fenced area or the area of a grazing permit

kéyah tálkáa'di dah si'ánígíí
island

kéyah táyi'déé' háá'á
island; literally land sticking up out of the water, pointed. (Alcatráz Island near San Francisco, California would be an example.)

kéyah dah naa'eeł
island: a large island in the ocean (the literal meaning is "land floating")

Una Vida 'Ruins' at Chaco Culture Park, New Mexico

-k'eh

English words with similar meaning:

place

-k'eh[10]
    can be translated as "place", but the basic meaning of -k'eh, refers to footprints, tracks, or trails, or areas used as gardens, or ruins, all of which show evidence of past activities of people or animals

shikéyahk'eh means "my place", my homestead, result of my activities.

anaasází t'óó bik'eh haz'ą́ refers to an abandoned Anasazi[11] place.

---

10 Linguists call this a locative enclitic.
11 Anasazi is the 'correct' Navajo term for the people who built Mesa Verde, Chaco, Aztec, etc. Archaeologists now tend to call them "Ancestral Pueblo", but most Navajo speakers would prefer "Anasazi".

tsé náázhoozh

Tsé náázhoozh, "a rockslide happened," at the north end of Table Mesa, New Mexico

English words with similar meaning:

rockslide

Tsé náázhoozh lá is a term for a rockslide that happened some time ago, and **tsé hiizhóósh** refers to a rockslide has just happened or is happening now.

Related terms:

tsé hoozh refers to small stones making a slide or a talus slope

łeezh nahaazhoozh refers to a slide of 'dirt' or soil

hashtł'ish náázhoozh refers to a mudslide

Section 6
Vegetation

hootso

English words with similar meaning:

green pasture or meadow

Hootso near Red Valley, Arizona

hootso refers to a meadow or pasture, usually natural.

dah náháltso refers to a small meadow or open grassy area, usually less than an acre.

Piñon-Juniper woodland

[some other terms for vegetation including trees, branches, and twigs]

hodílch'il refers to a tangle of vegetation, or a thicket, hard to walk through.

tsintah means in the woods (literally "amidst [or among] the trees").

tsįyi' means woods, or "in the heart of the forest". **Tsinyi'** has a similar meaning.

tsįyi'di means "in the woods or forest", and locates something in the woods. (**di** means "at that area").

Young and Morgan's Navajo dictionary says that **tsin** sometimes means "tree", but people say that the word is not used that way today. Now **tsin** usually means a stick, branch. Today, people would always specify the kind of tree when talking about trees.

Usually, areas filled with trees have different names, depending on the type of trees. Some examples of areas filled with trees are

- **chéch'iltah** is a term for among the oaks
- **gadtah** means among the cedars
- **nídíshchíí'tah** means in a pine forest, among the pines.
- **dówózhiitah** refers to a greasewood area
- **ts'ahtah** among the sagebrush
- **hoshtah** means among the thorny bushes
- **t'iistah** means among the cottonwoods

hahodishk'ą́ą́h means "I will burn out an area".

hahodeezk'ą́ą́' means burned in various areas but not everywhere;

hodichįįh refers to scorched ground where all plants are gone, where everything died.

Section 7
Earth Materials

A boulder, a lump of rock (stone)

tsé

English words with similar meaning:

rock (stone) [material]

tsétsoh: large rock, boulder (too large to lift up)

tsé nitsaa: large rock

tsézhin, chézhin: two different ways to pronounce the same word, meaning lava rock

tsé siziid: gravel spread out

tsé 'áwózí: river rock, pea gravel, terrace gravel, small and rounded

tsétah: among the rocks. In a rocky place.

tséta': between rocks, between the rock

tsé bee hodiwol: rocky area, area is strewn with rocks; gravel, boulders

tsé niit'ah and **tsé tsíín** both can refer to place near the base of a cliff. **tsé tsíín** refers when something is against the rock. Another related term is **tsé bichaha'oh**, which means 'in the shadow of the rock.'

tségháá': top of a rock, for example at a mesa or cliff

tsé biih hooldzis: small indentation in the cliff near the top of canyon wall, creating a pocket under the rock; also refers to a depression in the rock. Another way to say this is **tsé 'édzis**.

tsétát'ah: ledge or shelf on the face of a rock or cliff

tsé nesbaal: a large alcove in a cliff, similar to the ones at Mesa Verde

tsédáá': edge of a rock, cliff, precipice

tsék'iz: rock crevice, pocket or cleft within or between rocks; more like a fracture, but large enough to walk through

tséníí': hole or hollow in the rock, recess or niche (as in a canyon wall)

séí dah daask'id (sand dune)

séí

English words with similar meaning:

sand [material]

séítah
    sandy place, "among the sand"

séí dah daask'id
    sand dune

bis

English words with similar meaning:

clay, adobe [material]

bis

bis
    riverbank clay. Often translated as "adobe"

bistah
    clay area, "among the clay"

hashtł'ish

hashtł'ish
    mud; **hashtł'ish** refers only to wet mud. Dried mud is **bis**.

Picture taken near an old cabin in Red Valley, Arizona

łeezh

English words with similar meaning:

'dirt', dust, soil [material]

łeezh
    this form of the term łeezh is combined with other words to refer to dirt, dust, soil, etc.

łé-
    this is another form of łeezh

łeeshtah
    dirt floor, ground; this term is now used for a floor in general

dleesh
    white or gray clay (rhyolite tuff, kaolin)

łeezh dibahí
    fine dirt, dust

Index to Navajo-language Terms

'abikooh 22
'adah Hosh Łání 43
'adahiilí 9
'adahiilíní 9
'adzoogai 33
'ahéébíkooh 22
'ahibidiikooh 22
'ałts'ábíkooh 22
anaasází t'óó bik'eh haz'ą́ 65
'azką́ 53
'a'áán 45

baal 10
be'ek'id 11
bigiizh 39
bikáá' dah nímaz 55
bikooh 21–22, 24, 27, 29
bikooh bidáá' 22
biníi dah hastł'ah 30
biníi dastł'ah 30
biníi' dah hastł'ah 29
bis 81
bis dah 'azką́ 53
bistah 81

cháshk'eh 22, 27
cháshk'eh bidáá' 27
chéch'iltah 74

INDEX TO NAVAJO-LANGUAGE TERMS

chézhin 56, 77
ch'íbíkooh 22
ch'ínílį́ 13

dadzígai 33
dah 53
dah náháltso 71
dah nahwii 'eeł 17, 18
dah násk'id 51
dah yisk'id 51
dah 'ats'os 57
dah 'azką́ 53
dah 'iilį́ 9
dayílk'id 51
di 73
Diné Bikéyah 63, 64
dleesh 83
dówózhiitah 74
dzigai 33
dził 39, 49–50
dził binánii 50
dził dadiyinígíí 50
dził dah 'ats'os 50
dził diyinígíí 50
dził látah 50
dził nineel'ą́ 37
dził ní'á 50
Dził Yíjiin 49
dziłbą́ąh 50
dziłgháá' 50
dziłgiizh 39
dziłniit'aa 50
dziłtát'ah 50
dziłtł'ah 50

gadtah 74
-giizh 39

Háádahazts'aa' 43
haalk'id 51
haask'id 51
hahodeezk'ą́ą́' 74
hahodishk'ą́ą́h 74
hahoots'aa' 43
halgai 33
hashtł'ish 68, 81
hashtł'ish dits'idí 18
hashtł'ish nááźhoozh 68

INDEX TO NAVAJO-LANGUAGE TERMS | 87

hatsoh 22, 27
hats'ózí 22, 27
hazteel 35
hodichįįh 74
hodíłch'il 73
honíteel 35
hońteel 35
hoolts'aa' 45
hootso 71
hoshtah 74
hóteel 35

kéyah 63–64
kéyah dah naa'eeł 64
kéyah nineel'á 37
kéyah tálkáa'di dah si'ánígíí 64
kéyah táyi'déé' háá'á 64
-k'eh 65
k'iiz 41
-k'iz 41
k'iz 41

łé- 83
łeeshtah 83
łeezh 83
łeezh dibahí 83
łeezh nahaazhoozh 68

naanázkooh 22
nahasdzáán 61–62
náhodékaad 35
nahodits'ǫ' 17–18
náhooldzis 45
náhoolts'aa' 45
nástł'ah 29–30
ndzisgai 33
nídíshchíí'tah 74
nihosdzáán 74
-niik'iiz 41
ni' 37, 62
ni' bikáá' 62
ni'nineel'á 37

séí 79
séí dah daask'id 79
séí yáalk'id 52
séítah 79
shikéyah 64
shikéyahk'eh 65

88 | INDEX TO NAVAJO-LANGUAGE TERMS

tábąąh	4
tééh	15
tééh nílí	15
-teel	35
t'iistah	74
tó	3–4
tó adah 'iilį́	9
tó ahidahdiilį́	6
tó bidah 'iilį́	10
tó biníhííyį́	4
tó biyáázh	6
tó dah néigeeh	4
tó dah néigeehígíí	4
tó dah siyį́	4
tó dah 'iilį́	9–10
tó hadah dadeezlį́	9
tó hadah 'iilį́	10
tó hideezlį́	10
tó ndeezbaal	10
tó nílį́	5–6
tó nílínígíí	6
tó nineel'ą́	37
tó siyínígíí	4
tó yiląąh	6
tó 'ahidahdiilį́	6
tónteel	7–8
tónteel bibąąh	8
tooh	3–4
tooh nílínígíí	6
tooh siyínígíí	4
tsahtah	32
tsé	25
tsé bee hodiwol	77
tsé bichaha'oh	77
tsé biih hoodzáán	57
tsé biih hooldzis	78
tsé dah 'ats'os	57
tsé dah 'azką́	53, 57
tsé hiizhóósh	67
tsé hoozh	68
tsé k'aal	58
tsé łichíí' dah 'azkání	53
tsé náázhoozh	57, 67–68
tsé náázhoozh lá	67
tsé nané'á	57
tsé nesbaal	78
tsé níti'í	58
tsé nitsaa	77

INDEX TO NAVAJO-LANGUAGE TERMS

tsé siziid 77
tsé si'ą́ 57
tsé tsíín 77
tsé yaa hatsoh 58
tsé yík'áán 57
tsé 'áwózí 77
tsé 'édzis 78
tsé 'íí'á 55
tsé 'íí'áhí 55–56
tsé niit'ah 77
tsédáá' 78
tségháá' 78
tségháhoodzání 57
tségiizh 39
tsékooh 21, 23–24, 25, 27, 29
tsékooh bidáá' 23
tsék'iiz 41
tsék'iz 78
tséníí' 78
tséstł'ah haazt'i' 30
tsétah 77
tsétát'ah 78
tséta' 77
tsétsoh 77
tséyi' 25, 27, 57
tsézhin 55, 56, 77
tsézhin 'íí'áhí 55
tsin 73
tsintah 73
tsinyi' 73
tsístł'ah haazt'i' 30
tsjyi' 73
tsiyi'di 73
Tsoodził 49

yá 37
yá nineel'ą́ 37
yisk'iz 41

Index to English-language Terms

adobe [material] 81
alcove in a cliff, large 78
among the rocks 77
Anasazi 65
arch, rock 57
area, fenced 64
area, flat 35
area, open 35
area, rocky 77
arroyo 21–22, 27

bank (of a lake, river, or stream) 4
base (of a cliff) 77
base (of a mountain) 50
basin (natural water) 15, 45
bay 4
beach 4, 8
bed, stream 22, 27
bedrock 21, 23, 24
Bisti badlands 33
Black Mesa 49
blind canyon 29
body of water 3–4
body of water, large 3, 6
bog 17–18
boulder 77
box canyon 29
branches 73

bridge, natural 57
brink (of a rock canyon) 23
burn 74
burrow 45
butte 55, 57

canyon 13, 21–22, 23
Canyon de Chelly, AZ 5, 13, 25
canyon that converges 22
canyon, blind 29
canyon, box 29
canyon, curved 22
canyon, rock 23, 25
canyon, side 22, 23, 29
cascade 10
cedars 74
ceremonies 58
Chaco Canyon, NM 41
Chaco Culture Park, NM 65
Chaco Wash 15, 21
child 6
Chinle, AZ 13
Chuska Mountains 11
cistern 15
clay [material] 18, 83
clay area 81
clay riverbank 81
cleft (in the rocks) 41
cliff 29, 41, 58, 78
cliff, base of 77
cliff, indentation in 29
core, volcanic 55
cottonwoods 74
cove 29
Cove, AZ 33
crack 41
crater 43
crevice 41
crevice, narrow (in the rock) 39
crevice, rock 78
Crow Canyon, NM 23–24
cut 39

Defiance Plateau 49
depression 21–30, 45, 78
depression (large, concave) 45
depression (wide, shallow) 45
depression, longitudinal 22, 23

INDEX TO ENGLISH-LANGUAGE TERMS | 93

Dinétah 23, 27
dirt [material] 68, 83
dirt, fine 83
dry wash 22, 27
dune, sand 52, 79
dust [material] 83
dwellings, cliff 58

Earth World 61, 62
earth, surface of 21, 43, 45, 62
Earth, the 23, 25, 37
edge (of a dry wash) 27
edge (of a rock canyon) 22, 23
edge (of a cliff) 78
edge (of a precipice) 78
edge (of a rock) 78
edge, water's 4
El Malpais, NM 45

fenced area 64
flat area 35
flatland 33
flood 6
floor (dirt) 83
floor (in general) 22
foot (of a mountain) 50
footprints 65
forest 22, 73
forest, pine 74
Fruitland, NM 13, 29

gap 39, 41, 58
gap (in a mountain range) 39
gap (in a ridge) 39
gap (in a rock) 41, 50, 58
gap, narrow 22, 27, 30, 41
gardens 65
Grand Falls, AZ 9
gravel, pea 77
gravel, terrace 77
grazing permit 64
greasewood area 74
Great Salt Lake 7
ground 21, 37, 45, 62, 83
ground, scorched 74
gulch 21
gully 21, 27

harbor 4
hill 33, 51–52
hill, sand 52
hillside 29, 50
hogback 57
Hogback, The 17
hole 31, 45
hole (in the rock) 57, 78
hollow (in the rock) 78
hollow, big 58
homeland 23, 63
homestead 65
horizon 37
hunters 58
hunting 58

inlet 4
island 64

kaolin 83
Kayenta, AZ 25

lake 4, 45
lake, small 11
lake, very large 7, 35
land 33, 63–64
Largo Canyon, NM 23, 27, 33
lava rock 45, 55, 56, 77
ledge 50, 58
ledge on a cliff 78
Little Colorado River Canyon, AZ 23
Little Colorado River, AZ 9
Lybrook, NM 39

marsh 18
meadow 71
meadow (on a mountain) 50
mesa 53, 78
Mesa Verde, CO 58, 65, 78
Meteor Crater, AZ 43
Mexican Hat, UT 3
monolith 55
Mother Earth 61, 62
Mount Taylor 49
mountain 49
mountain pass 39
mountain range 50
mountains, sacred 50
mountainside 50

INDEX TO ENGLISH-LANGUAGE TERMS | 95

mountaintop 50
mud, wet 81
mud, dried 81
mudhole 17–18
mudslide 68

natural bridge 57
Navajo, NM 21
Newcomb, NM 15
niche (as in a canyon wall) 78

oaks 74
ocean 7, 35
offspring 6
open area 35
outcrops, rock 57
overhang, rock 58
overhanging ledge 58

Pacific (Ocean) 7
pass 39
pass, mountain 39
pasture, green 71
pea gravel 77
peak 50, 57
pine forest 74
pines 74
pit 15
place 65
place, rocky 77
place, sandy 79
plain 33
plateau 53
pond 4, 11
property, real 63

quicksand 17

rain 6, 21–22
range, mountain 39, 50
ravine 21
real property 63
recess, mountain 50
recess, curved 29
Red Mesa, AZ 53
Red Valley, AZ 53, 55, 71, 83
rhyolite 83
ridge, sharp 57
rim (of a rock canyon) 22, 23

river	6
riverbank clay	81
rock	10
rock (material)	57, 77
rock (overhanging ledge)	58
rock arch	57
rock canyon	23, 25
rock mesa	53
rock spire	55
rock, black	56
rock, dark	56
rock, hole in the	57, 78
rock, hollow in the	58, 78
rock, large	77
rock, perforated	57
rock, pointed	57
rock, river	77
rock, sitting	57
rock, vertical	10, 55, 58
rock, within the	25
rocks, among the	77
rocks, sloping	57
rockslide	67
rock trap	58
ruins	65
saddle	39
sagebrush	74
San Juan Chapter, NM	17
San Juan River	3, 15, 21
sand [material]	79
sand dune	52, 79
sand hill	52
Sanostee, NM	17
scorched ground	74
sea	7
seashore	8
shelf	50, 78
shelf on a cliff	78
Shiprock pinnacle	33, 58
Shiprock, NM	43, 53
shore	4
side (of a mountain)	50
skyline	37
slide of 'dirt' or soil	68
slope	50, 68
snow	6, 21
soil [material]	24, 68, 83
spire, lava	55

INDEX TO ENGLISH-LANGUAGE TERMS | 97

spire, rock 55
spring (of water) 6
stone [material] 77
stream 4, 5, 6
stream bed 22, 27
summit 50
summit (exact top) of a mountain 50

Table Mesa, NM 53, 67
talus slope 68
tangle of vegetation 73
terrace gravel 77
thicket 73
Thumb, The 55
Tóhdildoní Wash, NM 21
top (of a rock) 45
tracks 65
trail 65
trap, rock 58
trees 50, 73–74
trench, natural 21
tributary 6, 22
Tsegi Canyon, AZ 25
Tuba City, AZ 35
tuff 83
tunnel 45
twigs 73

Una Vida Ruins 65
Upper Fruitland 29

valley 15, 21, 22, 33
vegetation, tangle of 73

wash 15, 21
wash, dry 22, 27
water 33–18
water basin 15
water outlet 13
water, deep 15
water, running 5, 6
water, sitting 4, 11
waterfall 9–10
wildfire 35
Window Rock, AZ 29, 57
within the rock 25
Woodland, Piñon-Juniper 73
woods 73
World, the 61, 62